# Boreal Forests

**Patricia Miller-Schroeder**

**W**

**WEIGL PUBLISHERS INC.**

Published by Weigl Publishers Inc.
350 5th Avenue, Suite 3304, PMB 6G
New York, NY 10118-0069
USA

Web site: www.weigl.com

Library of Congress Cataloging-in-Publication Data

  Miller-Schroeder, Patricia.
  Boreal forests / Patricia Miller-Schroeder.
  p. cm. -- (Biomes)
  Includes index.
  ISBN 1-59036-345-0 (hard cover : alk. paper) — ISBN 1-59036-351-5 (soft cover : alk. paper)
1. Taiga ecology—Juvenile literature. I. Title. II. Biomes (Weigl Publishers)
QH541.5.T3M55 2005    577.3'7—dc22    2005005436

Printed in the United States of America
1 2 3 4 5 6 7 8 9 0  09 08 07 06 05

**Project Coordinators** Heather
C. Hudak, Heather Kissock

**Substantive Editor** Heather
C. Hudak

**Copy Editor** Tina
Schwartzenberger

**Designers** Warren Clark,
Janine Vangool

**Photo Researchers** Heather
C. Hudak, Kim Winiski

**Photograph Credits**

Every reasonable effort has been made to trace ownership and to obtain permission to reprint
copyright material. The publishers would be pleased to have any errors or omissions brought
to their attention so that they may be corrected in subsequent printings.

**Cover:** Getty Images/Stan Osolinski/Taxi (front); Getty Images/Michael Melford/National
Geographic (back left); Getty Images/Jeremy Woodhouse/Photodisc Blue (back middle); Getty
Images/Michael Lewis/National Geographic (back right).

**Getty Images:** pages 1 (Pal Hermansen/Stone), 3 (Michael Melford/National Geographic), 4
(Anders Blomqvist/Lonely Planet Images), 5 (Michael Lewis/National Geographic), 6 (Richard
Price/Taxi), 7 (James Gritz/Photodisc Red), 10 (Richard Nebesky/Robert Harding World
Imagery), 11 (Photodisc Collection/Photodisc Blue), 12 (Hans Strand/The Image Bank), 13
(Roine Magnusson/Stone), 14 (Paul Nicklen/National Geographic), 15 (Ron Chapple/The Image
Bank), 16 (Richard H. Johnston/Taxi), 17L (Jeremy Woodhouse/Photodisc Blue), 17R (Joseph
Van Os/The Image Bank), 18L (Alexandra Michaels/The Image Bank), 18R (David
Edwards/National Geographic), 19 (Stephen Sharnoff/National Geographic), 20L (Photodisc
Collection/Photodisc Blue), 20R (altrendo nature/Altrendo), 21L (Michael S. Quinton/National
Geographic), 21R (Michael S. Quinton/National Geographic), 22T (Randy Olson/National
Geographic), 22B (Digital Vision), 23L (Michael S. Quinton/National Geographic), 23R
(Heinrich van den Berg/The Image Bank), 24 (Photodisc Collection/Photodisc Blue), 25L
(Joseph Van Os/The Image Bank), 25R (Wilbur E. Garrett/National Geographic), 26 (David
Hiser/Stone), 27L (Roger Tully/Photographer's Choice), 27R (Bruce Forster/Photographer's
Choice), 28T (altrendo nature/Altrendo), 28B (Jacob Taposchaner/Taxi), 29T (altrendo
nature/Altrendo), 29B (Renee Lynn/Stone), 30 (Gary Vestal/The Image Bank).

**Cover description:** The boreal
forest biome can be found on
mountain ranges in many places
in the world.

All of the Internet URLs given
in the book were valid at the
time of publication. However,
due to the dynamic nature of
the Internet, some addresses
may have changed, or sites may
have ceased to exist since
publication. While the author
and publisher regret any
inconvenience this may cause
readers, no responsibility for any
such changes can be
accepted by either the
author or the publisher.

# CONTENTS

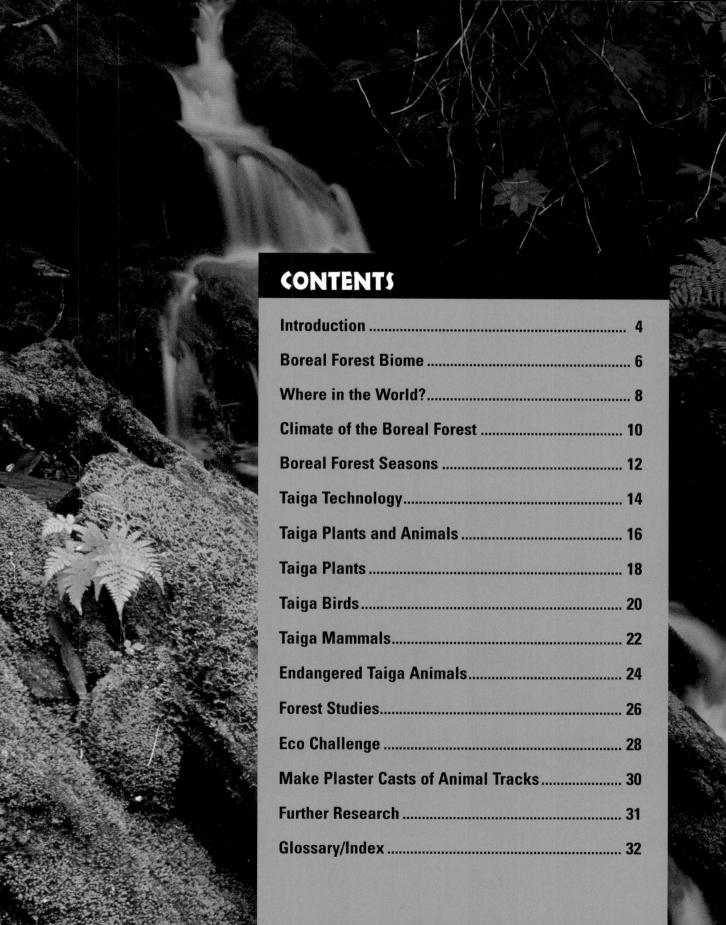

# Introduction

**E**arth is home to millions of different **organisms,** all of which have specific survival needs. These organisms rely on their environment, or the place where they live, for their survival. All plants and animals have relationships with their environment. They interact with the environment itself, as well as the other plants and animals within the environment. This interaction creates an **ecosystem.**

Different organisms have different needs. Not every animal can survive in extreme climates. Not all plants require the same amount of water. Earth is composed of many types of environments, each of which provides organisms with the living conditions they need to survive. Organisms with similar environmental needs form communities in areas that meet these needs. These areas are called biomes. A biome can have several ecosystems.

**Boreal forest rivers are fed by glaciers in mountains. Animals can use rivers in the boreal forest for travel, food, and as a source of water.**

The boreal forest is the largest biome on Earth. It stretches 50 million acres (20 million hectares) in an unbroken band across North America, Europe, and Asia. Cold north winds blow across the snow-covered land for much of the year. During the brief summer months, fires sometimes flare, destroying parts of the forest.

The boreal forest is home to many organisms that have adapted to this environment. Animals such as timber wolves, grizzly bear, lynx, moose, caribou, beaver, and voles live here. **Migratory** birds from warblers to whooping cranes spend their summers in the forest. Colorful mushrooms and **lichens** grow under the tall evergreen trees.

Within the boreal forest are two main types of forests. One is called "closed forest" because the trees grow close together. These forests are shady, and the forest floor is covered with velvety moss. The second type of boreal forest is called lichen woodland. The trees here are farther apart, leaving more open areas where lichens grow.

## FASCINATING FACTS

Taiga, boreal forest, northern forest, snow forest, and **coniferous** forest are different names for the same biome.

The word *boreal* comes from the Greek god, Boreas. He is the god of the north wind that blows through these forests.

Mushrooms play a key role in the boreal forest. Some mushrooms act as recyclers by decaying dead plant and animal matter. Others provide nutrition for trees. Many animals in the forest use mushrooms as a food source.

# Boreal Forest Biome

The boreal forest stretches across the northern continents of North America, Europe, and Asia, forming a circle. The forest's northern boundaries meet the treeless arctic plains, or tundra. This border parallels an imaginary line called an **isotherm.** Here, the average July temperature is higher than 50° Fahrenheit (10° Celsius). This temperature provides enough warmth in the summer for the trees to grow. The trees become smaller and more widely spaced apart as the forest extends farther north.

**Deciduous** forests and grasslands grow on the southern border of the boreal forest. This boundary parallels an isotherm where the average July temperature does not rise above 65° F (18° C). The evergreen trees of the boreal forest blend with the deciduous trees of the more southern forests.

**Many recreational activities, such as camping, hiking, bird-watching, and photography, can be enjoyed in the boreal forest.**

The North American boreal forest stretches across the continent from Labrador in the east to Alaska in the west. In some areas, this boreal forest extends more than 1,250 miles (2,000 kilometers) from north to south. Canada contains 24 percent of Earth's boreal forest. Eleven percent is located in the United States.

The European and Asian boreal forest stretches from Siberia in the east to Scandinavia in the west. In Asia, the boreal forest is 1,850 miles (3,000 km) from north to south at its widest point. Fifty-eight percent of Earth's boreal forest is in Russia. Norway, Sweden, and Finland together have 4 percent of Earth's total boreal forest, while Mongolia and China have 3 percent.

## FASCINATING FACTS

*Taiga* is a Russian word that means "marshy pine forest."

Together, the boreal forests of North America, Europe, and Asia cover 4.6 million square miles (12 million square km). This is larger than the total size of either the United States or Canada.

During the Ice Age about 18,000 years ago, much of the area that is now boreal forest was covered by glaciers up to 1 mile (1.6 km) thick.

Boreal forest snow is often soft and fluffy when it falls. A pail full of taiga snow will melt down to only 1 inch (2.5 centimeters) of water in the bottom of the pail.

Boreal forests contain bogs, fens, marshes, and lakes, as well as trees.

# WHERE IN THE WORLD?

**B**oreal forests are found in the Northern Hemisphere. This map shows where the world's major boreal forests are located. Find the place where you live on the map. Do you live close to a boreal forest? If not, which boreal forest areas are closest to you?

Arctic Ocean

North America

Atlantic Ocean

Pacific Ocean

South America

Boreal Forests

N

| 0 | 1000 | 2000 kilometers |
|---|------|------|
| 0 | 500 | 1000 miles |

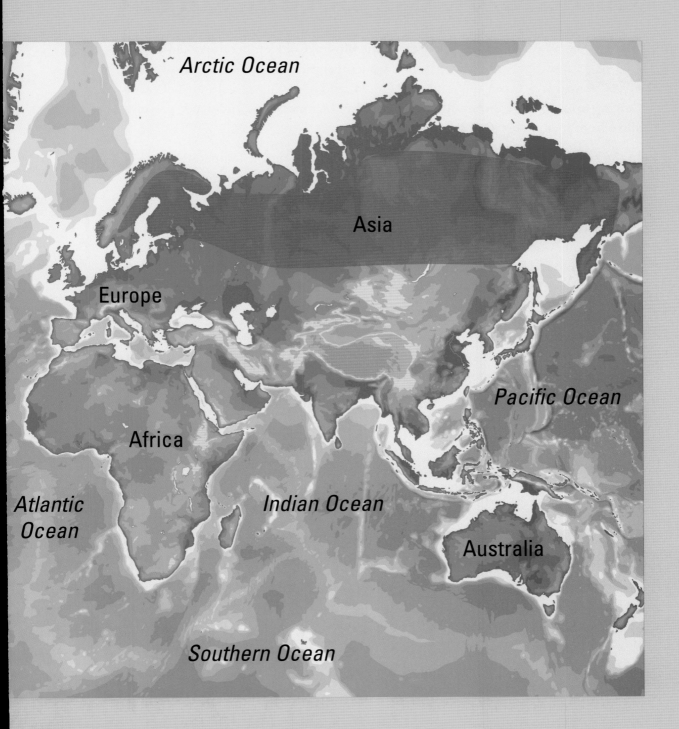

Arctic Ocean

Asia

Europe

Pacific Ocean

Africa

Atlantic
Ocean

Indian Ocean

Australia

Southern Ocean

# Climate of the Boreal Forest

The boreal forest lies below the arctic tundra. Cold temperatures can last from October to May. The average temperature for 6 to 8 months of the year is between −30° and −65° F (−34° and −54° C). Annual snowfall varies, ranging from 16 to 39 inches (40 to 100 cm). Cold temperatures and winds prevent snow from melting between storms.

In the boreal forest, summers are usually short, cool, and moist. Temperatures average between 20° and 70° F (−7° to 21° C). Some summer days can be hot and humid with temperatures rising above 80° F (27° C).

Precipitation falls as rain in summer and snow in winter. Cool temperatures slow **evaporation**. Much of the rain and melted snow is trapped in wetlands. In some places, the ground is permanently frozen. Many boreal forests have thick layers of moss that act as sponges to soak up water and keep the ground moist.

The slant of the Sun's rays during each season has a major effect on the amount of daylight that boreal forests receive. During winter, the nights are long and dark. There may be no sunlight for many days. In contrast, a summer day can have as many as 24 hours of daylight.

**Winter temperatures in the Tantra Mountains of Slovakia range from 22° to 26° F (−6° to −3° C).**

## FASCINATING FACTS

**In eastern Siberia, the average January temperature dips to −69° F (−56° C). At this temperature, exhaled breath freezes into ice crystals.**

**The coldest air temperatures in North America occur in the taiga. In January 1999, temperatures of −64° F (−53° C) were recorded near Fairbanks, Alaska.**

**Autumn is the shortest season in the boreal forest.**

## Permafrost

In some parts of the boreal forest, long periods of freezing weather cause permafrost. In these areas, the ground remains frozen for much of the year. However, the active layer, or top layer, may thaw, allowing plant roots to absorb water and grow. It is difficult for plants to grow long roots in such shallow soil. Their growth may be stunted, or they may fall over easily.

## Microclimates

Within the harsh climate of the boreal forests, there are many microclimates. These are small areas that are warmer or colder, wetter or drier, more or less shaded, or less windy than the norm. In many places, the forest floor is covered with fallen pine needles, twigs, and leaves that slowly decompose, creating new habitats for many insects and spiders. Loose bark and crevices on trees and fallen logs also provide shelter for insects, birds, and small mammals. Overhanging evergreen branches reduce the cold wind and provide pockets of warm shelter. Snow insulates many small mammals and insects throughout the harsh winter. Beneath deep snowdrifts, voles, mice, shrews, and a variety of insects and spiders live in warm tunnels and burrows. Some small plants can grow under the snow during winter.

# Boreal Forest Seasons

The Woodland, or Northern, Cree have lived in the boreal forest for centuries. The Woodland Cree have identified six seasons in the boreal forest. These are *sikwan* (spring), *mithoskamin* (break-up), *nipin* (summer), *takwakin* (autumn), *mikiskaw* (freeze-up), and *pipon* (winter).

Mikiskaw, the time of freeze-up, occurs between autumn and winter, usually in October. This is a time when the trees lose their leaves or needles and lakes are covered with ice.

Mithoskamin, the season of break-up, occurs from late March to late May or early June. During this time, snow melts, showing patches of bare ground. This season lasts until the thick lake ice melts. It can take a long time for lake ice to melt because it can be 3 feet (1 meter) thick. During mithoskamin, there are long hours of daylight, but little moisture is released from frozen ground and lakes. Warm winds, along with 12 to 16 hours of sunshine each day, dry out the trees. Fires often occur during mithoskamin.

## FASCINATING FACTS

One year after a fire in Alaska, 2,000 spruce, 500 poplar, and 800 birch seedlings per acre (0.40 hectare) were counted.

Jack pines depend on forest fires to distribute most of their seeds. Their pine cones are sealed with a resin that only opens in intense heat. Jack pine seeds can withstand temperatures of 1,300° F (700° C).

The cone shape of a coniferous tree allows snow to fall when it gets too heavy for the branches. This prevents the branches from breaking.

## Fire and Regeneration

In most years, thousands of fires break out across the boreal forest. Lightning strikes cause one-third of the forest fires. The remaining two-thirds of forest fires are started by careless people. The average lightning fire is ten times larger than fires caused by people. Each part of the forest will burn at least once every 150 years.

The forest provides a large amount of fuel for fires. Conifer needles and branches are highly flammable, especially during dry weather. Lichens burn easily and grow on many tree branches.

New growth begins almost immediately after a fire. Plants sprout through the ash-covered soil. The wind blows plant seeds to the burnt area. Other plants, such as fireweed, thistle, willow, birch, and aspen, lie dormant in the soil. Some tree species have roots living underground. Birch trees grow new sprouts around the burnt rim of trunks. Heat from fires opens jack pine cones, spilling seeds on the forest floor. As the forest grows, trees become taller, blocking out sunlight. Trees that enjoy shade, such as the spruce and balsam fir, may become the most common species in the boreal forest.

**The smoke from forest fires can travel great distances, even to other continents.**

# Taiga Technology

**O**ne hundred years ago, few people ventured into the vast boreal forest. However, the Woodland Cree adapted to the harsh living conditions. They made snowshoes and sleighs to travel across the land. They also built snow houses called *quin-zhee*. To make a quin-zhee, the Cree shoveled snow into a pile and let it harden for at least 1 hour. Then, they burrowed into the snow, making a hollow, cave-like space. Inside the quin-zhee, body heat and **insulating** snow warmed the temperature to 25° F (−4° C)—warmer than the temperature outdoors.

Today, new technologies help scientists navigate the boreal forest and learn about its plants, animals, and ecosystems. Scientists often study samples from forest areas to learn about different organisms and how many species live there. They use live traps and nets to collect birds and animals. Scientists record the age, gender, condition, and weight of each animal. They also mark animals with identification tags or bands. Scientists use the data they collect to understand how ecosystems change over time.

**Electronically tagging an animal allows scientists to monitor its movement throughout the animal's entire life.**

Scientists also use radio telemetry, radio tagging, or radio tracking. A transmitter attached to an animal sends a signal to a receiver. Scientists use the radio signal to track the animal's movements. This method of tracking provides a great deal of information about an animal's habitat, territory, migration, activity, and life history.

Remote satellite sensing uses satellite images to show what types of trees grow in an area, where wetlands and lakes occur, where fires are burning, the effects of climate change, and areas affected by human development. Remote satellite sensing helps scientists maintain healthy forests. Scientists can also use remote satellite sensing to monitor, map, and model forest fires. **Infrared** satellite images show burning vegetation. They show the location of active forest fires, how the fire is behaving or moving, and the size of the area burned.

The Global Positioning System (GPS) uses satellite technology to find exact locations any place on Earth. People use GPS to navigate the boreal forest.

## FASCINATING FACTS

If the sky is cloudy, forest fires might not be detected with remote sensing monitors.

Using GPS is like giving every square foot (0.1 square meter) on Earth a unique address.

# TAIGA PLANTS AND ANIMALS

**D**espite its harsh climate, the boreal forest is home to many life forms. However, there are fewer species here than in southern forests. As winter approaches, many bird species, some butterflies, salmon, and mammals such as caribou migrate to warmer climates. In the spring, birds return to the boreal forest to enjoy the abundant food that swarms of insects and the lush forest vegetation provide. Some hardy bird species remain in the boreal forest throughout the year.

## PLANTS

Evergreen trees, such as pine, spruce, and balsam fir, are characteristic of the boreal forest's coniferous trees. These trees keep their needle-like leaves year-round. They provide food and shelter for birds and animals that remain in the forest during the harsh winter. Some conifers, such as tamarack and larch, drop their needles. Other hardy deciduous trees of the boreal forest include poplars, aspens, and birch. Although trees are the most visible plants in the boreal forest, there are many other types of vegetation. Berry-loaded bushes, such as blueberries and

Many animals feed on blueberries.

cranberries, and other shrubs, including willows and alders, grow in many areas. Velvety mosses carpet the forest floor, and lichens grow on and between many trees. Fungi, such as mushrooms, grow alongside small plants, including the Indian pipe and the pitcher plant.

## BIRDS

During the short spring and summer months, the boreal forest is alive with birds. Few species remain throughout the entire year. In the warm months, birds such as warblers and whooping cranes return to the forest from as far away as South America. Waterfowl such as ducks, geese, swans, loons, shorebirds, gulls, and herons live in the forest during spring. Goshawks, eagles, ospreys, great horned owls, and great grey owls are a few of the **raptors** that frequent this biome.

The goshawk is a large, carnivorous bird.

Pine martens belong to the weasel family.

## MAMMALS

In the boreal forest, mammals range from the shrew to the grizzly bear and moose. **Herbivores** include moose, elk, caribou, deer, snowshoe hare, beaver, muskrat, mice, and voles. Many carnivores use a variety of hunting strategies to prey upon herbivores in the boreal forest. Wolverines hunt alone, while wolves stalk their prey in groups. Other hunters include coyote, fox, black bear, grizzly bear, ermine, least weasel, marten, otter, lynx, cougar, and shrews. Many of the same mammals live in the boreal forests of North America, Europe, and Asia.

# Taiga Plants

## Coniferous Trees

Most boreal-forest trees are coniferous evergreens. Conifers keep their green leaves for more than one season, allowing the tree to begin **photosynthesis** early in spring. Conifers have needle-shaped leaves with a waxy coating. These leaves lose less water in spring and summer than broad leaves. Most evergreens lose their needles after 2 or 3 years. Some, such as spruce, keep their leaves for as long as 8 or 9 years. Conifers, such as larch and tamarack, lose their leaves every year.

The bark of an aspen tree is edible.

## Deciduous Trees

Some boreal forest trees are deciduous. These trees lose their leaves in autumn and re-grow them in the spring. This allows the trees to use less energy in the winter months. Fewer branches break from the buildup of snow and ice as well. Most of these deciduous trees are broadleaf, such as birch, aspen, and poplar, or shrubs, such as willow, alder, and blueberry.

## Fungi

Many types of fungi grow in the boreal forest. Fungi often grow near other plants. Mushrooms are fungi that come in a variety of shapes, sizes, and colors. They grow in and on living and dead trees and branches. Mycorrhizal fungi grow on the roots of conifer trees. These fungi look like thin hairs. They help tree roots absorb moisture and minerals to make food.

Spruce trees are common in North America.

## Lichens

Lichens form from algae and fungus. Lichens come in a variety of shapes and sizes. They live on the ground, as well as on logs, rocks, and tree branches. Some lichens are important food sources for caribou and reindeer. These lichens are called reindeer mosses.

## Moss

Moss is a small, feathery plant that grows on the forest floor and in bogs. In cool, moist places with acidic soil, moss can grow several feet (1 to 2 m) thick. Moss can absorb huge amounts of moisture. Sphagnum moss can hold up to twenty times its weight in water.

White-tailed deer feed on witch's hair lichen.

### FASCINATING FACTS

Some conifers reproduce by layering. If a lower branch touches the ground and soil covers it, the branch can form roots and develop into a new tree. The new tree is a clone, or exact duplicate, of the parent tree.

Taiga trees harden during autumn. Hardening enables them to withstand temperatures as cold as –40° F (–40° C).

Aspen tree clones that share the same root system are considered one organism. They can stretch more than 100 acres (40 ha) and have thousands of tree trunks.

Many small, carnivorous plants, such as pitcher plants, sundews, and bladderworts, live in the boreal forest. These plants lure insects using color, odor, and nectar. They trap insects using suction cups, sticky flypaper, or pitfall traps.

# Taiga Birds

## Migrant Birds

During the winter, the boreal-forest bird population drops from about 300 species to twenty or thirty types of birds. In the most northern regions, only two or three species remain. Most migrant birds fall into two groups. The first group travels short distances. In North America, these birds migrate to the southern United States. Birds in this group include bald and golden eagles, great blue herons, loons, kingfishers, tree swallows, warblers, sparrows, and hawks. The second group of birds travels longer distances to tropical climates such as Mexico and Central and South America. This group includes ospreys, peregrine falcons, cuckoos, martins, redstarts, tanagers, sparrows, swallows, and warblers. In Europe and Asia, long-distance migratory birds travel to Africa or other tropical destinations.

Black-capped chickadees communicate using at least fifteen different calls.

## Scatter Hoarding

Many year-round bird species store food for the winter. Gray jays begin hoarding in June. They store hundreds of food items, including insects, spiders, berries, and mushrooms. Jays pack the food into pellets and coat them with saliva. Then, they cram the pellets into cracks in the bark of a tree or in a cluster of conifer needles. Chickadees and tits are also hoarders. These birds store seeds, berries, and insects. Each bird hides thousands of food items each day in needle clusters, lichens, bark, curled leaves, and broken branches. Some chickadees use silk from spider webs and cocoons to hold the seeds in place.

An osprey's wingspan is about 5 feet (1.5 m).

## Non-migrant Birds

Some bird species remain in the boreal forest year-round. North American birds in this group include goshawks, grouse, ravens, nuthatches, kinglets, jays, woodpeckers, chickadees, crossbills, and grosbeaks. Many owls, including great-horned, great gray, boreal, barred, and hawk, are year-round residents, too.

## Burrowing

Many small birds, including chickadees, redpolls, sparrows, and snow buntings, burrow under the snow. Using its feet, wings, and beak, a Swedish willow tit can dig an 8-inch (20-cm) tunnel under the snow. Grouse and ptarmigan also burrow under the snow. To shelter from wind and cold, boreal-forest birds often roost in the shelter of thick conifer branches or squeeze into tight tree cavities.

**The white-tailed ptarmigan is the smallest of all ptarmigans.**

## FASCINATING FACTS

To save energy during long, cold nights, some birds can lower their internal body temperature. Chickadees and Siberian tits lower their body temperature at night from 104° F (40° C) to 86° F (30° C).

Gray jays nest from late February to early March, when temperatures are as cold as –20° F (–29° C). They lay eggs before April.

# Taiga Mammals

## Dwellings

Most mammals live in protected microclimates. They build dens, dig burrows, or create tents under conifer branches. Crevices and holes in trees and logs also provide shelter. Deep snow provides insulation. In these shelters, the temperature reaches 25° F (−4° C) when the weather is −40° F (−40° C) outdoors. Some mammals, such as mice and voles, huddle together in burrows during the winter. Some mammals build year-round dwellings. Beavers build dams made of logs, branches, and rocks that they seal with mud from lakes and ponds. Much of the dam is underwater. Beavers cut vast supplies of wood to feed on during the winter, and store fat in their tails during the summer and autumn.

The beaver is North America's largest rodent.

A grizzly bear spends up to 8 months in its den.

## Hibernation

Some mammals gain large amounts of weight in autumn. These animals **hibernate** throughout the winter. Both black and grizzly bears enter their dens in late autumn and do not leave until spring. During this time, bears maintain a state of deep sleep. Their heart rate slows, and their body temperature drops.

To escape predators, snowshoe hares will run up to 27 miles (44 km) per hour.

## Camouflage

Some boreal-forest animals change the color and texture of their fur coats depending on the season. Snowshoe hares turn white in winter and brown in summer. This helps them blend in with the snow or soil. Their winter coats contain clear, white hairs that have air pockets inside. These air pockets act as extra insulation against the cold. Hares also have large, furry feet that act as snowshoes, allowing the animals to run over the surface of the snow without sinking. The lynx, which is a main **predator** of the snowshoe hare, also has large, furry feet. A lynx's feet help the lynx chase prey across the snow.

## FASCINATING FACTS

Bears usually eat between 5,000 and 8,000 calories a day. Humans eat between 2,000 and 3,000 calories a day. In late summer and autumn, bears eat 15,000 to 20,000 calories a day. During this time, they gain about 1.5 to 2.2 pounds (0.7 to 1 kilograms) per day.

Wolverines are also called skunk bears, devil bears, and gluttons. A large male weighs about 60 pounds (27 kg). Wolverines can tackle prey as large as moose and caribou.

Unlike other species of squirrels, tree squirrels have no cheek pouches. They sleep during the day in holes in trees or in nests of leaves.

# Endangered Taiga Animals

nimals in danger of becoming extinct are classified as endangered. This means there are so few of the species alive that they need protection to survive. The grizzly bear, Siberian tiger, pine marten, whooping crane, and peregrine falcon are just a few of the boreal forest's animals that are considered to be endangered. In the United States, it is illegal to hunt or harm endangered animals.

Pollution traps large amounts of **greenhouse gases** in Earth's atmosphere. The global climate is currently warming ten times faster than at the end of the last ice age. Many scientists believe this warming trend will continue, permanently changing the boreal forest. Snow may melt earlier in the season, summers may become drier, forest fires may occur more frequently, and permafrost may melt. The southern part of the boreal forest may disappear completely, becoming parkland and arid grassland. It will be difficult for plant and animal species to adjust and survive.

**Destruction of coniferous forests and hunting for pelts are the biggest threats to the American pine marten.**

Clear-cutting, a logging technique in which all of the trees in an area are removed at once, destroys habitat and disrupts normal forest growth. Plant and animal species lose their natural habitats. Overharvesting of animals through commercial hunting and fishing can cause problems for certain species. Recreational vehicles and the roads built for them disrupt habitats, breeding and nesting areas, and migrating pathways.

**Pesticides such as DDT have been a major threat to the peregrine falcon.**

Hydroelectric dams built on major rivers change the water flow, flooding thousands of acres (hectares) of land. This destroys habitats and migration routes for salmon and caribou. Large numbers of wildlife drown or become displaced. The James Bay Hydroelectric Project in Quebec, Canada, has caused the drowning deaths of thousands of migrating caribou.

**Boreal forests are experiencing deforestation as rapidly as rain forests.**

# FOREST STUDIES

**F**rom working with animals or plants in the boreal forest to researching the effect of climate change in boreal forest ecosystems, most forest-related jobs require a background in biology, ecology, or environmental studies. Choosing a career in wildlife conservation, biology, or forest management is ideal for people interested in ecosystems and animals.

## WILDLIFE BIOLOGIST

- Duties: studies wildlife and their environments

- Education: bachelor's, master's, or doctoral degree in biology, zoology, environmental studies, or ecology

- Interests: biology, the environment, conservation, science, animals, plants, ecosystems

Wildlife biologists enjoy learning about animals and how different species relate to each other and their environments. They study wildlife, prepare environmental projects, perform field research, analyze data, inventory plant and animal communities, and practice environmental impact studies. Wildlife biologists also prepare information in brochures, books, and slide shows for presentations to schools and other groups.

## ECOLOGIST

- Duties: studies the relationship of living things to each other and to their environments

- Education: bachelor's, master's, or doctoral degree in environmental science or biology

- Interests: animal behavior, animal and plant communities, how ecosystems work, environmental issues

Ecologists specialize in animals or plants and their interactions. Ecologists examine endangered species or ecosystems. They study the effects of pollution and develop species recovery plans. Some study soils or climate.

## FORESTER

- Duties: works to ensure people use forests wisely and limit harm to forest habitat and wildlife

- Education: bachelor's, master's, or doctoral degree in forestry

- Interests: forest ecosystems, land use, habitat restoration, wildlife conservation, botany

Foresters work for government agencies and forestry companies as researchers and consultants. They plan forest use, forest-renewal projects, and advise about appropriate forest-management techniques. They assess the effects of pollution, human activities, and forest fires on forest habitats and wildlife.

# ECO CHALLENGE

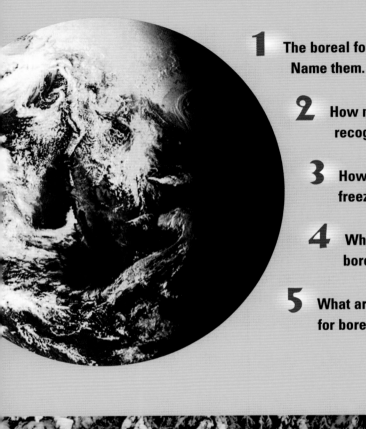

**1** The boreal forest appears on how many continents? Name them.

**2** How many seasons do the Woodland Cree recognize in the boreal forest?

**3** How many months have temperatures below freezing in the boreal forest?

**4** Which country contains the largest boreal forest area?

**5** What are three other names for boreal forest?

**6** How are most forest fires started?

**7** What do most boreal-forest birds do in winter?

**8** Name four endangered boreal-forest animals.

**9** What are the most common trees in the boreal forest?

**10** What attracts so many birds to the boreal forest each spring?

# MAKE PLASTER CASTS OF ANIMAL TRACKS

**B**oreal-forest animals leave their footprints or tracks in the snow and on the wet ground. You can learn how to identify the tracks of different animals and then try to make your own in plaster.

## MATERIALS
- pencil
- paper
- plaster of Paris
- disposable plastic container

1. Explore your community to find an interesting footprint. Sketch the print using a pencil and paper.

2. Mix a batch of plaster of Paris.

3. Pour the plaster into the plastic container.

4. If you have a pet dog or cat, gently press the animal's paw into the plaster. If you do not have a dog or cat, press your bare hand or bare foot into the plaster. Wait for it to harden.

5. Check your yard for animal prints. If you find one, fill it with plaster. Once the plaster hardens, you can lift it from the print. Can you tell what sort of animal made the print? Is it from a cat or dog? Could it be from another animal, such as a deer?

# FURTHER RESEARCH

**H**ow can I find more information about ecosystems, boreal forests, and wildlife?

- Libraries have many interesting books about ecosystems, boreal forests, and wildlife.

- Science centers, museums, and interpretive programs at zoos and parks are great places to learn about ecosystems, taiga, and wildlife.

- The Internet offers some great Web sites dedicated to ecosystems, taiga, and animals.

## BOOKS

Drake, Jane and Ann Love. *Cool Woods: A Trip around the World's Boreal Forest*. Toronto, ON: Tundra Books, 2003.

Henry, J. David. *Canada's Boreal Forest*. Washington, DC: Smithsonian Institution, 2002.

Lynch, Wayne. *The Great Northern Kingdom: Life in the Boreal Forest*. Markham, ON: Fitzhenry & Whiteside, 2001.

## WEB SITES

**Where can I learn more about forests and animals?**

Boreal Forest Network
www.borealnet.org

**How can I learn about current forest concerns?**

Global Forest Watch
www.globalforestwatch.org

**Where can I learn more about protecting boreal forest areas?**

The Taiga Rescue Network
www.rescue.org

# GLOSSARY

**coniferous:** trees that have cones and needles

**deciduous:** trees that lose their leaves at the end of the growing season

**ecosystem:** a community of living things sharing an environment

**evaporation:** the process of changing from a liquid or solid to a gas

**greenhouse gases:** atmospheric gases that reflect heat back to Earth

**herbivores:** plant-eating animals

**hibernate:** to spend winter resting or sleeping to save energy

**infrared:** invisible wavelengths

**insulating:** keeping heat in

**isotherm:** a line drawn on a map linking areas of the same climate

**lichens:** organisms made up of algae and fungi that grow on tree trunks, rocks, and the ground

**migratory:** move from one area to another

**organisms:** individual life forms

**photosynthesis:** the process in which a green plant uses sunlight to change water and carbon dioxide into food for itself

**predator:** an animal that hunts and kills other animals to eat

**raptors:** birds of prey

# INDEX